"Nat can pick a dog,"
said Mom.
Nat picks a big, black dog.

1

Nat grins at the black dog.

Nat's dog is Zack.

Zack is Nat's best friend.

Can Zack help Nat spell? No!

Nat can toss a stick.

Zack gets the stick fast.

"I am off!" said Nat.

"You can not go," he said
to Zack.

5

"Zack ran off," said Dad.

"Can I get him back?"

"Help me get the dog back,"
said Dad.
"Zack! Zack!" Mom said.

"Nat is back," said Dad.

"Zack is the best dog,"
Mom said.

The End